Christianity's HOAX
Jesus the Winemaker

AL & ALICE LUNDEN

WESTBOW
PRESS
A DIVISION OF THOMAS NELSON

WestBow Press books may be ordered through booksellers or by contacting:

WestBow Press
A Division of Thomas Nelson
1663 Liberty Drive
Bloomington, IN 47403
www.westbowpress.com
1-(866) 928-1240

ISBN: 978-1-4497-9625-9 (sc)

Library of Congress Control Number: 2013909254

Printed in the United States of America.

WestBow Press rev. date: 7/10/2013

Table of Contents

Acknowledgements

The writers wish to especially acknowledge the Inspiration, Study, and Writings of the many modern men and women who have discovered the Amazing and Astonishing Divine and Absolute Truth that Jesus was most certainly NOT a maker or user of intoxicating wine and that His Religious Heritage through Moses, would also not have tolerated the use of this or any other communal or recreational drug. We also appreciate the work of many members of the staff of West Bow Press who have assisted in the publication of this book.

Our utmost and most vital Appreciation, of course, is to Jesus!

Introduction

An Appeal for Holiness!

Please read this book! It is being written only to expose the apparent myth or false belief that Jesus was a maker and user of wine rather than a maker of grape juice. That myth, which has undermined Christianity for 17 centuries, appears to have perhaps come from a consensual agreement or compromise between the early Roman Christian Church and Constantine the Great to cease Christian execution and persecution if his Army Generals and Political Leaders were given membership and key positions of leadership in the new Holy Roman Catholic Church.

Compromise might have come at the time in Roman history when religious persecution had advanced to bestiality when even women and children were being killed and eaten by hungry lions and dogs. Such an agreement of Judeo-Christianity with Rome would certainly not allow or abet continuation of the Mosaic Doctrines of prohibition of things like communion wine and social drinking. A much later but very similar compromise is the apparent consensual agreement of several Protestant churches to perhaps also avoid controversy by using wine for communion although some inspired clergymen must be aware of this deception.

Do not be deceived by inaccurate Bible translations or by misguided teachers who thus actually deny the absolute Holiness of Christ by acceptance of those false translations and not from the inerrant and infallible Hebrew Bible and the Greek Testament. This book is really being written solely to defend Jesus' Holiness and to expose the blasphemy and sacrilege of all false or misleading translations of the Holy Bible. Let us simply strive to live and love one another in the TRUE IMAGE of Jesus! HE is the Only Man who has ever lived or shall ever live on earth without even a trace of unholiness!

Kudos to Our Beloved Lord and Savior, Jesus Christ!

EVIDENCE WHICH
DEMANDS A VERDICT

Was Jesus "Guilty" or "Not Guilty" of making wine as His very first of many miracles? Please examine the evidence and come to your own conclusion on this very vital issue! We firmly believe that Jesus was NOT a maker of exotic intoxicating high-point wine but a maker of far more nutritious and delicious grape juice! Christians should perhaps get back to the basics and study the original Hebrew Bible and the Greek Testament! The World must get to know the real Jesus and not fake man-made imitations of the Lord who has been created by perhaps well-meaning but misguided human translators of the Bible.

Christianity's Image and the World Image of "The

Man Without Sin" does not appear to be the true image of Who Jesus really was! Agape' Love? Unconditional forgiveness? Infinite wisdom? Faithfulness? Fellowship? Gentleness? Honor? Honesty? Absolute separation from sin? None question these attributes of Christ. But what about HIS HOLY IMAGE of abstinence from the use of wine or other forms of alcohol and other mind-altering drugs? From beer, ale, champagne, a little scotch, or a bit of brandy at bedtime? We do so believe because of the writings of a host of extremely gifted and inspired Christian Bible scholars.

We do not even suggest that it is a cardinal sin to use any or all of the above but only that it is wrong to presume to drink them "In the Image of Christ" since HE actually made no wine. The Hebrew word yayin, the Greek word oinos, and even the Latin word vinum are generic words for all grape beverages which must always be studied in context to identify either wine or grape juice.

Let us then just get back to the basics. Consider our Judeo-Christian Heritage. Our walk down an aisle at the supermarket would reveal bottles of Jewish sounding (kosher?) Manischeweitz or Mogen-David wines. But would Moses really have authorized God's Divine Blessing on these fermented and alcoholic wines? Supposedly not, from the original writings of Moses in the Hebrew Bible, although the Latin Vulgate, the King James Bible and the New International Version do not agree. Would Jesus have honored or would HE have dishonored Moses' sacred restrictions against leavened bread and wine?

Discussion or preaching against winemaking or alcohol consumption are pretty much avoided in many or most Christian churches! Why? Possibly because most churches assume, infer, or absolutely insist that Jesus did make wine and that they thus must also dogmatically defend their long-time church traditions and doctrines? This is much like the fairy tale about the scam of the crooked tailor with non-existent thread who tricked his Emperor into believing that he was making a beautiful new suit. The tailor succeeded until one little boy was horrified when he saw his beloved Emperor in his underwear.

Perhaps most churches also choose to simply ignore the obvious destructive use of alcohol, excessive social drinking, or alcohol addictions of some or many of their members? Just twenty months ago I and my wife wrote the book, Jesus the Winemaker: Satan's Most Effective Lie, and had it published by West Bow Press, a Division of Thomas Nelson.

This more comprehensive book was written primarily for Christians and for Christian churches and intended for sale in Christian Bookstores and in major Bookstores. It was published in November 2011 and is available at Barnes & Noble.

We have since realized that this sacrilegious and blasphemous lie is not promoted only by Satan but perhaps mostly by Christianity. The truth, the whole truth, and nothing but the truth can be found only in the God-Inspired words of the Hebrew Bible and in the Greek Testament which are absolutely infallible and inerrant. All other Bible translations are subject to the personal beliefs or objectives of those who have "authorized" and have thereby perhaps also paid for their translation or publication.

Consider especially the Latin Vulgate and the King James Version of the Bible. These two long-time and world famous Bibles were written and published, respectively, by Rome and The Holy Roman Catholic Church and by King James and his Anglican Church. Their writers would most certainly have been restricted and controlled by the specific beliefs of Catholicism and Rome and by the social mores, moral standards, scriptrual interpretations, vested interests, and personal whims of King James I and the Church of England.

We also realize that writing only to those Christians or others who buy books at shopping malls, at Christian book stores, or at exclusive booksellers like Barnes and Noble is much like preaching mostly to those in our beautiful and expensive churches who live in equally expensive and often lavish mansions, houses, and apartments. We are ignoring the masses who live on our streets, in crowded trailer parks, in old houses and apartment buildings; who work very long tedious hours in factories, fields, forests, foundries, offices, stores or in other modern "salt mines"; and who often seek and find only false fellowships through the use and abuse of some form of the demeaning drink which made Jesus famous!

Many of these men, women, and children are counted among the millions of Americans with no real homes or with dysfunctional or broken homes because of the use or abuse of wine or other drugs. Statistics are not available but Christian homes are very susceptible to alcohol use and abuse, spousal abuse, sexual sin, separation, divorce, child abuse, etc. But most Christians in America and in the world are certainly also not aware that Jesus Christ was actually a maker and user of grape JUICE and that, in His absolute Holiness, HE apparently did not EVER use any intoxicants!

Please now just consider the massive EVIDENCE from Scripture, Logic, Science and Modern Medicine which support the hypothesis that Jesus was NOT a maker or user of wine. Remember that Christianity is not Assembly of God, Baptist, Catholic, Episcopal, Lutheran, Mennonite, Methodist,

Nazarene, Presbyterian or any other but is truly non-denominational!

Something appears to be amiss or out of balance in wisdom or spirituality when mature adults in any society profess great faith in a Supernatural Being but subject themselves to the use of substances which even control or destroy their mental and physical abilities! All Christians must simply study all of the evidence of Scripture, Divine logic, absolute truth, science and modern medicine and then find Jesus innocent or guilty of making wine! HE cannot be both the Drug Czar of all Creation and the reigning King of Christian Love, Sobriety, and Responsibility.

What is your personal verdict? Guilty or not guilty?

Part 2

SCRIPTURAL EVIDENCE
WHICH DEMANDS A VERDICT

Please just consider the Scriptural Evidence which would need to be presented at a Jury Trial or at a Supreme Court Trial to carefully and painstakingly determine: 1-If Jesus made grape juice? or 2-If HE had actually made intoxicating wine as His very first Miracle at that wedding reception in Cana? What evidence would those jurors or judges need to hear to determine Jesus' guilt or innocence of the apparent sin of bringing five jugs of wine to a small town wedding including scores of children, teenagers and pregnant women?

How many realize that there were actually two very different kinds of wine in the Hebrew Bible and in the Greek Testament? Strong's huge 1340 page

Exhaustive Concordance and his Hebrew-Chaldee and Greek Dictionaries are very thorough in their compilation and discussions of the many words which pertain to grapes and to wine and to the making of wine. All these words should or must be read and studied very thoroughly both in their original Hebrew and Greek languages and in those original contexts of good or evil to determine if they only refer to fresh grapes or grape juice or to fermented or intoxicating wine.

Strong's and Zodhiates Numerical Listings of all Key Words of scripture along with any parallel Bible allows even those with little or no knowledge of Hebrew or Greek languages to study those writings in their original contexts. The contrasts between God's blessings of grapes and unfermented grape juice and the evils associated with intoxicating wine appear to be very evident. Strong's Concordance includes about 150 references to the evils of wine and about fifty references to the overwhelming blessings of grapes or grape juice.

Several inspired and learned men and at least one very special lady have researched and published their findings to support the Holiness of Jesus and to expose the absolute evil of wine but Alice and I have no personal claims to this divination of Scripture. They and many others richly deserve serious consideration for the Nobel Prize in Religion for their discoveries. These saintly workers include Samuele Bacchiocchi,

William Patton, Donald Stamps, James Strong, Robert Teachout, Muriel Van Loh, and Spiros Zodhiates. These Christian pastors, workers, or lay persons are from many different denominations and are all great writers.

Exactly what did each and all of these workers discover in the original Hebrew and Greek scriptures? The undeniable truth that each and every Bible reference to grape juice or wine included clear contextual evidence of either good or evil. That over 150 references related to the undesirable use or abuse of wine and that over fifty related to desirable qualities of grape juice or grapes. This is evident from the Book of Genesis through the Book of Revelation with absolutely no grey areas of confusion or ambiguity.

Consider the first mention of wine in Genesis when Noah got drunk soon after getting off the Ark. The pre-flood greenhouse climate and the leaven or yeast-free environment before the flood was obviously a sterile place where fruit juices did not undergo fermentation or putrefaction into wine or vinegar. Then the sun emerged through the perennial blanket of water vapor when that very first rainbow was seen by Noah and his family!

Moses was told to scrupulously remove all traces of leaven from every home in Israel just before the Passover meal. This was apparently considered to be absolutely necessary so that the bread and grape juice would not become impure from any contamination. We feel that this is just like the absolute Holiness of Jesus. His grape juice was not like the alcohol free beer of today from which most of the ethanol has been removed but which still does contain some traces of this drug. John the Baptist did not even drink grape juice, perhaps or presumably so that others would not suspect that he was drinking wine or so that others could not intentionally serve him any wine.

Jesus did not even accept the evil of hard cider while in dire thirst on the cross. And Paul's advice to Timothy was not to use a little wine but to use a little grape juice for his health problem, which appears to have been either chronic acidity or ulcers.

The scriptural evidence certainly does appear to warrant thorough study and consideration. We also sincerely believe that Jesus has been found guilty of the alleged evils or crimes of making, serving, and drinking wine and of encouraging the use of wine during marriage. This evidence suggests that the use and abuse of this drug, especially within Christian marriages, should be considered to be a violation against humanity, against Christianity, and against the sacred institution of marriage!

'We blaspheme Jesus is we accuse Him of making or using wine!

Part 3

DIVINE LOGIC AND ABSOLUTE TRUTH

Consider first, of course, the claim that Jesus was and is both a man and a God of absolute Holiness! The jurors and judges would need to determine and define good and evil. Could an intoxicant like wine ever be considered to not be evil? Is intoxication evil? Just what acts, attitudes, or actions might logically be considered to be good and desirable or evil and undesirable? Would it especially be good or evil to provide an intoxicating drug to the many children and women present at that Cana wedding reception? Did Jesus realize that the bride or other women at the wedding would or could give birth to children with physical or mental disabilities if they drank wine? Of course HE knew this although it took mankind nearly 2,000 years to discover the curse of Fetal Alcohol Syndrome which has devastated or destroyed millions since the time of Noah and of Lot who was twice seduced by wine.

150 GALLONS of intoxicating wine? Good grief! Or should we say "Give me a break!" as there is nothing good about the availability or use of too much wine? Why would Jesus make so much more wine if they had simply run out of wine after they had already been drinking for several hours? Unless this grape juice was like a pick-up load of watermelons at a church picnic so that everyone could eat or drink and then take a melon or a portion home to share this Divine Provenance with those who had to stay home to watch the sheep and feed the chickens. To further spread the Great News that the long-promised Messiah had begun His earthly ministry in Galilee. Divine truth cautions against temptation.

Divine Logic would also need to consider the undeniable world devastation which is attributable to the close association of drug use with Christianity. Why is this so? Simply because most Christian children have been brought up to firmly believe that Jesus really did make wine so that they see little or no wrong in the use of this or other drugs! My old World Book Encyclopedia credits or blames Christianity with the introduction of grapes to southern Europe and the development of the massive wine industries of Italy and France. These two nations are among the world's worst in frequency of Fetal Alcohol Syndrome.

Consider the Cana Wedding as the model for Christian weddings, weddings of important personages, and essentially all world weddings for the past seventeen centuries. We have no doubt whatever that all three of our beloved daughters and their bridegrooms were under the carnal influence of alcohol and not under the sacred presence and power of the Holy Spirit on their wedding nights. Peer pressure for alcohol use at all weddings is both from within the church and from their Christian or non-Christian friends. We have little or no doubt that those three tragic marriages would probably have been filled with Christian love rather than absolutely doomed for destruction by the use and abuse of alcohol or other drugs if we and they had then known of Jesus' abstinence.

How many of our brides or grooms experience alcohol induced consensual sex, suffer from spousal abuse, or are actually victims of the equivalence of rape on their wedding nights because of the use or abuse of alcohol? Or during their courtship? Consider courtship: What does consensual sex really mean when alcohol is involved? Date rape and acquaintance rape are absolutely not limited to the male species! And the aspect of virginity is very different for the sexes! We strongly suspect that promiscuous business or working women, women on college campuses or in night clubs, dinner clubs, or bars have even more virginal conquests of men than many predatory males. Many of those evil men or women are actually guilty of intentional rape.

The precious wedding night should be the most awesome, memorable, and sacred night of every Christian marriage with absolute mental and physical acuity! Jesus' Holy Spirit should reign supreme; should provide the absolute tenderness, understanding, and sensitivity of unconditional and agape' love; and should so vibrantly etch this Precious Time into the memories of husband and wife that "Their Wedding Night" will truly survive throughout eternity.

Agape' or unconditional love is the Divine Logic and Absolute Truth of God's Divine Plan for Christian Marriage!

Part 4

SCIENCE AND MODERN MEDICINE

Most Christians do not even consider the undeniable scientific fact that the ethyl alcohol or ethanol in wine actually is the most commonly used addictive drug in the world. Ethanol is also, of course, both an intoxicant and a deadly poison! These vital scientific facts are actually very evident in both the Hebrew Bible and in the Greek Testament but not in the Latin Vulgate, in the King James Version, or in other translations of the Holy Bible. We sincerely believe that this absolutely is the most significant factor in the widespread use and abuse of alcohol within Christianity.

Science and medicine really do provide a wealth of evidence of Christianity's Hoax which is nearly as convincing as the concrete evidences of Scripture and of logic! Consider the poisonous nature of ethanol or ethyl alcohol and the indisputable fact that it IS a highly intoxicating, very toxic, and mind-altering drug. Ethanol is very easy to ferment, brew or manufacture and is now a household word through its prominence in the fuel industry. Wine and any of the many other forms of alcohol are also the most common entry drugs into the horrible and destructive worlds of all other even more addictive drug abuses.

Consider our bodily responses to this drug. Alcohol moves quickly into the bloodstream from drinking or oral consumption and then directly into the brain. Inhalation of ethanol, which is a new fad, gives even more rapid entry into the bloodstream through the lungs. This will, of course, give a much quicker feeling of euphoria or "high" from the immediate absorption of ethanol through the tremendous surface area of the air sacs in the lungs and thence into the brain than from normal absorption through the digestive system. This, of course, quickly results in rapid brain dysfunction and then to mental, visual or physical impairment.

Modern Medicine especially provides even more proof positive with news and discovery of the worldwide tragedies of Fetal Alcohol Syndrome! FAS is especially damaging for Native Americans and for pregnant women and their children in France and in Italy. Jesus simply could not, by the nature of His Absolute Holiness, have used or encouraged the use of such a poisonous substance!

I did not realize just how destructive this chemical was until I began to wash out my paint brushes with only 10% ethanol gasoline several years ago. The skin on my hands quickly became very cracked and broken from the desiccating effects of this fuel although this had never before happened for me with regular gasoline. Identical tissue desiccation injury is true for the mouths and throats of those fools who presume to be Real Men when they consume even small quantities of straight whiskey, vodka, grain alcohol, scotch, or moonshine. The toxic and intoxicating ingredient in wine or in hard apple cider, beer or ale or in a nefariously or secretly spiked wedding punchbowl is, of course, the same chemical.

Medical and Scientific Research has also proven that the ethanol or alcohol which goes into the bloodstream causes permanent injury to brain tissues as well as to the cells, tissues and organs in the respiratory system, the digestive tract and the circulatory system. Our hearts, lungs, kidneys, and stomach are especially sensitive. Only time will tell what new lung and brain damage might result from ethanol inhalation. Science has also shown that permanent mental injury results from long-time or excessive drinking of alcohol although such claims are denied by many or most users.

We suspect that many or perhaps most repeat DUI offenders are not physiologically able to recognize their mental and physical limitations and legal restrictions after a few drinks so they absolutely need other physical or mental restraints. One especially expensive but often necessary restraint is long-term incarceration. But there is a Better Way! Please just consider the Presence and the Power of the Holy Spirit of Jesus Christ in the human heart, mind, body, and Soul?

HE and His Holy Spirit brought me freedom from alcohol use at age 50 when I surrendered my will to Him and discovered HIS Abstinence! HE shall forever be my escape hatch from alcohol use! I was no longer a maker and user although I had been a long-time user and abuser of alcohol. I had literally no effective control over the abuse of alcohol until that time and have enjoyed over thirty years of absolute sobriety since then! We cannot aspire to live "In the Image of Christ" if we don't have full knowledge and understanding of that image.

Part 5

CHRISTIAN MARRIAGE AND OUR AMERICAN SOCIETY

We feel that Christianity's Hoax of Jesus' wine is especially evident in the breakdown of family values and traditions throughout Christian society in America! Studies have shown that divorce rates are nearly equal for Christians and for non-Christians. Why? Perhaps partly because wine and beer are not really even thought about as drugs. We suspect that alcohol use and misuse is perhaps even the primary cause of the undeniable and deplorable nationwide and worldwide breakdowns of Christian marriages and families. Consider the overwhelming and Godly importance of family heritages throughout the Hebrew Bible and the tremendous breakdown of such heritages in the America of today! We are a nation of broken families in which single-parent families or separated parents are most common. Few children have simple multi-generational family trees tracing back to their ancestors--- Because of separations and divorces.

Consider just three simple aspects of Christian Marriage and of the Christian Family in regard to the mores and customs of Our Society as they relate to alcohol:

A- Holiness and Morality.

B- Financial Responsibility.

C- Christian Witness.

We especially suspect that alcohol has been the world's most effective cause or promoter of sexual sin and un-holiness since Noah emerged from the Ark and for many centuries of and within Christianity. We suspect that the false modern Image of Jesus as a maker and user of intoxicating wine has made that possible. We feel that money spent on alcohol, bad choices related to alcohol use and abuse, and the horrible consequences of those poor choices blaspheme both Christianity and Jesus. We also especially believe that the false Judean Heritage of Judeo-Christianity and the authenticity of the Holy Bible are being blasphemed by this Hoax of Christianity.

Part 5-A: Holiness and Morality: We believe that a vast majority of all Christian weddings occur in the false Image of another Jesus through the worldwide belief that HE actually did make real wine. This includes weddings in those churches which actually use grape juice for communion but honor the false inerrancy and infallibility of the NIV and King James Bibles, the Latin Vulgate, or other Bibles. Those churches clearly also pay lip service to American society's complete acceptance of social drinking. We actually thought we were best advised or even obliged to teach Our Children to be responsible users of alcohol rather than to even suggest that their best choice would be total abstinence. They became alcohol users "Just like Dad" and their first marriages with and to alcohol abusers all ended in the pain of abuse, separation, and divorce.

We erred miserably because our children were all victimized by the alcohol and drug use and abuse of both our and their friends, families, and spouses. I am in tears more than 30 years later as I write this but their church weddings or receptions were followed by their wedding dances which left little opportunity for them to perhaps even remain sober for the most sacred, tender, wonderful, and memorable experience of Christian marriage. We firmly believe that many such sincerely and well-intended intended "Wedding Day Benefits" might often or usually actually be very detrimental to the sacred institution of marriage. We Praise God that our wonderful and special small and semi-private church wedding included only three friends, the pastor, and his wife with no wine, champagne, beer, or liquor.

ALL marriages should be based on lifetime bondings like those of Adam & Eve, Noah & his wife, Abram & Sarai, Boaz & Ruth, and even like that of David & Bathsheba. We especially consider the undoubtedly lifetime vow of faithfulness of that honored bride and groom at the world's very first genuine "Christian" wedding at Cana when Jesus first announced His Earthly Ministry as the long-promised Messiah! What about Jesus' role in that most sacred Cana Wedding? Would HE really allow and abet the bride or groom to have anything less than absolute sobriety, total mental and physical acuity, and ultimate or maximal spiritual awareness on their wedding day or night? We think not.

How much family dissention and discord can be blamed on the evils of alcohol use? Just picture a Utopian-Christian marriage, family, and society or even a Utopian and non-Christian wedding, marriage, family and society in which there is absolutely no use or abuse of alcohol or other drugs! Dysfunctional families are, of course, especially evident when either or both marriage partners or even their children have any inherited affinity or weaknesses for alcohol abuse or addiction. My step-mother brought my father out of a terrible shadowland of alcohol addiction but still encouraged our social drinking in the false Image of Jesus rather than to encourage our abstinence.

Please just consider Christianity's pursuit of holiness in the true and absolute sinless Image of Jesus! None question His Agape' or unconditional love but many passionately deny His absolute abstinence from the use of alcohol or other drugs. We must blaspheme His Holy Name!

Part 5-B Financial Responsibility: How does financial responsibility relate to love and happiness within marriages and to the agape' love and stability of Christian families? Consider the enormous weekly or monthly expenditures related in any way to the use or abuse of alcohol for any typical Christian family? Compare this amount with their contributions for church and for other charitable offerings or with some designated percentage tithe to their church. What are other factors of financial or personal responsibility, especially for any who may be troubled with the malady of alcohol abuse? Many Christians are blessed with overwhelming generosity but such generosity can actually overflow in many negative ways while they are under the influence of alcohol or other drugs. Irresponsibility, sexual immorality, gambling, unwise investments, exorbitant or unholy spending and money wasted on alcohol are prime examples.

What factors of financial or personal irresponsibility regarding alcohol use and abuse abound in our society? Many abusers of alcohol or other drugs in jails or in prisons? Alcoholics living on our streets or in homeless shelters, without employment or on welfare because of alcohol related felonies, poverty, highway or other accidents, or disabilities? Why is our nation the world's leader in incarceration? Bankruptcies? Gambling Abuse? Spousal or Child Abuse? The list is endless! Many of these ills relate to the use and abuse of wine, beer, and other alcoholic drugs in our society and within Christianity and from the influence of these drugs.

The alcohol in sweet wines, such as those used for communion in many Christian churches rather than like the grape juice used by Moses and Jesus, APPEARS to be both nutritious and delicious, but is instead intoxicating and destructive. Communion Wine is like a beautiful, good, and delicious sip of sweet nectar to our youth which gradually or suddenly turns into an ugly, evil, and bitter serpent of lust. Wine is also, of course, precursor to alcoholism for many of us who are genetically susceptible to alcohol addiction or misuse from this devastating disease! The bottom line is that it is a mind-altering drug which does not impart responsibility but brings on confusion and irresponsibility.

We must also especially consider the financial implications of alcohol abuses in our state and federal government, in our judiciary, in education, in business and industry, in the military and even within our Christian Churches? How can we thus encourage Godly, moral, and responsible practices and policies in all of these areas? We may never know how many of the widely publicized cases of sexual misconduct within Christian leadership are related to the use and abuse of wine but we do know that these leaders have ready access to intoxicating communion wine.

Part 5-C Christian Witness

Christian Witness is actually the primary responsibility of each and every true Christian. This is especially evident and especially crucial for all of us who presume to be Christian parents, grandparents siblings, uncles, and aunts. We must simply be living examples of the vital Image of Jesus so that it is imperative that we know very precisely just Who and What HE was and Who HE is. That vital knowledge and that Holiness Image can only be discovered and attained by careful reading and study of Scripture and other Divine writings, by extensive fellowship with other genuine Christians, and by hearing the Word of God spoken in our churches, in our cathedrals and in all other places of worship.

The Christian Witness of those who practice total abstinence is just like that of John the Baptist. We believe that he apparently chose to drink absolutely no wine or grape juice so that none would ever even suspect that he might be drinking wine. Consider the "Christian Witness" of those who drink only a small amount of alcohol and are always "in control" but might well encourage irresponsible persons to drink. Consider those who presume to be "makers of wine in the Image of Christ" but whose wine is given or sold to others who are genetically or physiologically susceptible to alcohol misuse or abuse?

All children are essentially and totally dependent upon their parents, much as we Christians are totally dependent upon Almighty God for everything, until they are old enough to know Who God really is so our "Christ Image" or "God Image" is very important. I was really and especially a living Image of a maker, user, and abuser of wine when our children were very small, when they were in elementary school, and when they were in High School. Our church also provided a very clear and vivid living Image of Jesus as a maker of wine at the World's Very First Christian Wedding and as a user of wine for Holy Communion!

Drunkenness has perhaps always been a prime topic of stage humor for theatre, movies, plays, and television but it is absolutely tragic for the men, women, and children in prisons who have committed all sorts of crimes when under the influence of alcohol! I have been in a prison visitation program for about 25 years where I regularly visit complete strangers who have been incarcerated for various crimes. Most of my prison friends have been jailed for alcohol related offenses. My current prison friend is 55 but will not be eligible for parole until he is 65. His crime was trying to steal beer and money from the clerk in a convenience store when he was drunk about ten years ago.

Christian, which is not his real name, grew up in a Christian but wine drinking family just like mine. Nearly all Christian families in America have Christian friends, neighbors, or family members who are in or have been in prison for various crimes related to alcohol use or abuse. Many others are victims or survivors of horrible auto accidents which have been caused by drunk or drinking drivers. How does Christian witness really relate to the use or abuse of wine or other drugs? I can certainly and easily lead a friend, son, or daughter into their lifetime dependence upon alcohol even if or when I have complete control over my personal actions.

I was an obsessive maker of wine for a number of years. I turned the tiny atom-bomb scare and cold-war age fallout shelter in our basement into a wine cellar. At first I was really very particular about the quality of the fruit which I used; about time and temperature; and about proper filtering, bottling, and aging. This was an inexpensive way to make really good quality wine. I sought clarity, body, proper aging and all other quality factors of great wine but then something happened. I just sought the end result. I began to make wine from almost anything and everything and cared only for the alcohol! I used poor quality fruit, wheat, and even beets when nothing else was handy and paid little attention to quality, handling, or aging and I eagerly drank the settlings and yeast along with the wine with no aging whatever. My Christian witness was of no priority.

What especially are the "Christian Witness" values of parents like Alice and I and of Christian churches which actually promote the use of wine or other alcoholic drugs by presuming that Jesus probably or actually was or even might have been a maker of wine just because human or mortal church leaders and Bible translators say that HE made wine at Cana? How do attitudes or implications of moderation, sobriety, tolerance, or abstinence in our Christian society relate to the TRUE Image of Jesus? Was HE responsible for and then perhaps indeed guilty or innocent of turning water into wine? We each must ask and then strive to answer this vital question with Christian conscience.

We choose to believe there will be jubilation in

Heaven if and when millions of worldwide Christians suddenly discover that Jesus was an abstainer, avoid alcohol abuse, and live in sobriety! And there is certainly nothing to lose if we believe this and if we are wrong.

Our Christian Witness is or could easily be one of absolute abstinence and sobriety "just like Jesus" and like that of John the Baptist.

Part 6

KINGDOM LIVING AND SPIRITUAL BAPTISM

Consider the GREAT NEWS of Both God's Holy Kingdom and Jesus' Earthly Kingdom! These are: 1-The Eternal Kingdom of Almighty God In Heaven and 2-The Earthly Kingdom of Jesus! Living in the Earthly Kingdom is to live completely under the Presence and Power of Jesus' Holy Spirit. HE gives us Unconditional Love, Peace, Joy, Comfort (which really is the ability to tolerate every mortal hardship-such as overwhelming pain, sorrow and debility), and countless other gifts such as our absolute assurance of the Promise of eternal life!

Jesus' Kingdom is actually not only the wonderful promise of the "pie in the sky by and by" which is to live "Eternally with Christ" but it is also "Living in

Peace on Earth" until we are called Home. Consider our earnest petitions, from The Lord's Prayer, for "Peace on Earth as it is in Heaven". God's Peace is an Awesome Gift! It is Jesus' Agape' Love, Patience, Wisdom, Joy, Kindness, and Gentleness. God's Peace is completely beyond our human or mortal understanding.

We really begin LIVING IN "The Earthly Kingdom of Jesus" on this awesome Earth of His Creation when we are "Baptized in and with the Holy Spirit!" The Presence and Power of the Holy Spirit lifted me out of the fears, fantasies and agonies of my then troubled life at 3:15 Friday afternoon July 31, 1981 and I have lived on a Mt. Everest plateau of love, peace, and joy for over 31 years. Baptism of the Spirit is neither sprinkling or immersion or coming to the altar at a Billy Graham Crusade! It is, however,. Total Commitment of Your Life to the Triune God!

Baptism of the Holy Spirit is indescribable and may be a thousand different things for a thousand different people so I can only speak for myself. It was then and still is outpouring of blinding tears of love, joy, peace, and comfort! It was cleansing of my subconscious mind of evil memories, unwanted desires, and fantasies. It was the purging of hatred. It was a hunger to know everything about Jesus and to strive to live In His Image. It brings our immediate repentance and our tears of remorse for the many involuntary and voluntary or premeditated sins which always will be a part of our humanness.

LORD! JESUS! We Love YOU so much!

Part 7

The Last Chapter and Personal Testimonies

I want to share an interesting observation. Some people, like Alice, sometimes, often or always begin reading books by first reading the last pages, paragraphs, or chapters. This part of our book is especially for those who, even when reading suspense mysteries or adventure stories, have that strange quirk, habit, or addiction. I also have some special questions for those who profess to be Christians as well as for those who do not. If you do so profess, have you really given or dedicated your heart, mind, and soul to Christ? If you are not a Christian or if you are only a carnal Christian, as I was until age 50, will you consider future living as a Spiritual Christian by earnestly striving to live in the True Image of Christ?

This book has been mostly about the True Image of Jesus and Christianity and is mostly about patterning the Image of precisely Who HE was and of Who HE is! Alcohol abuse is only one small facet of sinfulness but can be an enormous roadblock to Christian witness for those who are essentially wired for overindulgence. I was Both a sincere Christian and an alcohol abuser mostly because I firmly believed that Jesus was a maker and user of wine. I was "Baptized and Confirmed in and by The Church" when I was young but was not "Baptized by the Holy Spirit" until my personal and private "conversion" at the age of 50.

WWJD? We again ask what, indeed, WOULD Jesus do? Those four wonderful words became my motto at 3:15 on that afternoon of July 31, 1981 and have now been my "Marching Orders" every day for over thirty years. But how does or can anyone know exactly what Jesus would do? Scholars ask "What and where shall I study?" and then "Where shall I work?" Young men and women ask "Whom shall I date and whom shall I marry?" and then "Where shall we live?" What, indeed, would Jesus do in your situation and want you to do in that situation?

But none can really answer any of these vital questions without thorough knowledge of the personal and Divine nature of Jesus and without an intimate personal relationship with Him! "Living In Jesus'

Image" is, of course, impossible but I and Alice and all other Christians can and must strive, pray, read, study, and fellowship together to strive toward that impossible goal: Jesus' True Image! His unconditional or Agape' Love, which was and is His Primary Gift of Spiritual Baptism to all true believers, was best evidenced by His voluntary acceptance of torture and death on the cross!

His Forgiveness of those who placed Him on the cross or engineered His rejection, His suffering and His death! His Patience with men like me who flounder around in the mental confusion of carnality or sin simply because we worship false man-made Images. Jesus' gentleness, empathy, and sympathy for all are without equal. We absolutely MUST begin to accept Jesus' infinite holiness and remove this albatross of His long-standing and false recognition as the Drug Czar of America and of the World rather than the Reigning King of Creation!

Consider my personal use and abuse of alcohol dating back to our Graduate School years at the University of Florida. My real Christian Mentor and Graduate Study Advisor was a renowned church and community leader with a well-stocked liquor cabinet. Social drinking was then a natural part of Christianity and we had absolutely no clue that Jesus was not a maker and user of wine. We knew absolutely no students or faculty members who did not drink alcohol regularly in our church's false and contrived Image of Jesus. But I and some members of my family have genetic susceptibility to alcohol use for entertainment and for alcohol abuse when things go wrong.

My alcohol use or abuse was, of course, only a weekend ritual as I was much too busy with my research work and my course work to even consider any drinking during the week. Our three wonderful children were all born in those four years of intensive research study, church, Bible Studies, and church fellowship. We were all extremely devout in worship as both my wife and I had always been and would be for the rest of our lives. This pattern continued while I was a Research Scientist in Oak Ridge, Tennessee and finally a University Professor for many years in the Midwest. I thus maintain that Judeo-Christian Doctrines and church traditions clearly embrace alcohol use and abuse and tolerate other aspects of carnal Christianity.

I and we were Sunday School teachers for many years and were always very staunch witnesses for all of those church doctrines and traditions which included things like tithing, regular attendance and midweek worship. We did not, however, teach our children abstinence from wine or other alcoholic drugs in the true Image of Jesus but attempted to teach them to just be responsible users of wine in the church's false Image of Christ. Scripture tells us to avoid temptations which lead to sin but the social or legal use and misuse of alcoholic drugs provide a perfect oxymoronic example of Satanic influence over Christianity.

What happened? Our daughters naturally became users and abusers of alcohol in the real image of their father and in the false Image of Jesus. Tragedy came when they all dated and married alcohol users and abusers who were also irresponsible and unfaithful. Abuse! Desertion! Separation! Divorce! Children suffer the most from the sins of their parents. I was then a frustrated Associate Professor in teaching and research on the campus of So. Dakota State University but we are all victims of ignorance and unholiness

I had been a Christian since infant Baptism but only a carnal Christian until the age of fifty. Twice daily I escaped from our coffee-room banter and sacrilege, sought refuge in the Gospel of John, and spent those precious 15-minute interludes with Jesus. Then I surrendered to Jesus at 3:15 Friday afternoon on July 31, 1981 in the sanctuary of my small but private office!

Wow! Heaven absolutely erupted in that little room with my overwhelming tears of love and joy as I gave my heart, mind, body and soul to Jesus! I had never known such peace! Christian love, joy and peace are not just the highest peaks of a roller coaster of hills and valleys but the plateau at the top of the world!

The contrast between carnal Christianity and Spiritual Christianity is as wide as the void between Heaven and Hell! Jesus has now given me comatose and untroubled sleep with only pleasant dreams and visions for an amazing 32 YEARS! I had been troubled by nightmares, evil dreams, and macabre visions since the untimely death of my mother when I was 12. HE has thus cleansed my subconscious mind of all evil and unwanted memories. HE has awakened me, without the need or use for an alarm clock, for many thousands of duties, obligations, and opportunities to love Him and to serve Him in many countless ways during that time.

Jesus has given me survival, healing, and excellent health and vigor through two massive heart attacks and by-pass surgeries at the ages of 54 and 69 and from acute appendicitis at the age of 80. HE has also given me His Miracles of Protection and avoidance of death or severe injury from at least five seemingly unavoidable and potentially fatal accidents. HE healed Alice through miraculous back surgery and HE has given us nearly 60 years of overwhelming happiness and togetherness in our marriage. Kudos to Our Father, to His Only Begotten Son and to The Holy Spirit! Our God is an Awesome God! He is Alive!

Selected References for Christianity's Hoax: Jesus the Wine Maker

Bacchiocchi, Samuele, Wine in the Bible, Biblical Perspectives # 8, Andrews University, 4990 Appian Way, Berrien Springs, MI 49103, 2004.

Lunden, Al & Alice, Jesus the Winemaker: Satan's Most Effective Lie, West Bow Press, 2011.

Patton, William, Bible Wines, Star Bible, Fort Worth, TX 56118, 1874 (1981 Edition).

Stamps, Donald C., The Full-Life Study Bible, Zondervan Corp., 1992.

Strong, James, Strong's Exhaustive Concordance of the Bible, Abingdon Press, 1890 (1973 Edition).

Strong, James, Concise Dictionaries of the Hebrew Bible and the Greek Testament, Abingdon Press, 1890 (1973 Edition.)

Teachout, Robert P., The Use of Wine in the Old Testament, Doctoral Dissertation, Dallas Theological Seminary, 1979.

Teachout, Robert P., Wine, The Biblical Imperative: Total Abstinence, Trinity Baptist Church, Taylor, MI 48180.

Van Loh, Muriel, Alcohol in the Bible, Promise House Publ., Canistota, SD 57102, 1992.

Zodhiates, Spiros, The Hebrew-Greek Key Word Study Bible, AMG International, 1984 (1991 Edition).

Zodhiates, Spiros, Lexical Aids to the Old and New Testaments, AMG International, 1984 (1991Edition).

www.ingramcontent.com/pod-product-compliance
Lightning Source LLC
Chambersburg PA
CBHW050337290526
45785CB00006B/2529